THE WATCHMAKER

ROBERT J RANI

EDITED BY

MIKE RANI SHERI RANI

ISBN-13: 978-1-7349626-0-4

Publisher: King and Lord Publishing

Library of Congress Control Number:
2020908259

of Zondervan Publishing House."

Cover image: Andrey Armyagov
cookelma's gallery, Moscow, Russia
© /Stockfresh.com

Cover cross image: #995505 christian cross
by drizzd © /Stockfresh.com

This book is dedicated to:

My Lord and Savior, Jesus Christ.

"Seek me and live."
Amos 5:4

And also to Mike and Sheri Rani for all the hours spent editing for me, for Him, and for you.

PREFACE

"If you tell a big enough lie and keep repeating it, people will eventually come to believe it."

Although the source of that quote is in dispute, if you believe what Jesus tells us in John 8:44, then you will see that the initial source must be the devil...

"You belong to your father, the devil, and you want to carry out your father's desires. He was a murderer from the beginning, not holding to the truth, for there is no truth in him. When he lies, he speaks his native language, for he is a liar and the father of lies."

Accepting a lie as truth can never make it the truth.

Contents

THE WATCHMAKER

The battle between intelligent design and the theory of evolution has been in the news and the courts and will probably continue until Jesus returns.

Evolutionists claim that the intelligent design movement is merely a pretense to introduce the religious belief of biblical creationism into our public schools. The real reason they oppose the teaching of intelligent design is that it leads one to the inevitable conclusion that if man did not evolve, then there must be a creator.

† † †

In 1859, Charles Darwin said in his *On The Origin of Species*, the proof of evolution would be the "transitional" or "intermediate" fossils, but he also lamented over the lack of evidence:

"Why then is not every geological formation and every stratum full of such intermediate links? Geology assuredly does not reveal any such finely graduated organic chain; and this, perhaps, is the most obvious and gravest objection which can be urged against my theory."[1]

If a species were to evolve into another, there would have to be transitional fossils which would have characteristics of both the former and the latter. Although there are between 100 and 200 million fossils in museums[2] there is not one verifiable transitional fossil.

In the year 2000, while I was at the Museum of Natural History in New York, I asked where the transitional fossils could be found, but they did not have any. The only displays of transitional forms were artist's conceptions. Just like the theory of evolution they were all man made. One of those on display was of archaeoraptor, which was touted as a link between dinosaurs and birds. The missing link had been found. Well, not really, it was later determined that it was actually part dinosaur and part bird which had been glued together by a Chinese farmer.[3]

This was not an isolated incident. In Sussex, England in 1912, Piltdown Man was declared to be a link between ape and man. In 1953 it was determined to be a forgery. The skull was that of a man and the jaw was from an orangutan.[4] The teeth had been filed down and it had been stained to give it the appearance of being older than it was.

In 1922 Nebraska Man was promoted as another supposed link between ape and man. The artists conceptions in *The Illustrated London News* were based on what had been found, which was one tooth.[5] The tooth turned out to be of a peccary (a type of pig.) The missing link is still missing. There are no transitional fossils because there has been no evolution, but don't take my word for it. The following quote is from Dr. D. James Kennedy of Coral Ridge Ministries:

"In the last few years, notable scientists like Steven Jay Gould of Harvard (now deceased) and Niles Eldredge of the American Museum of Natural History (both evolutionists) and others, have said that it has long been a trade secret of paleontology[I] that there are no intermediate links. Now the time has come to tell the public: There are no intermediate links. I have in my library copies of letters from the leading museums in the world: the British Museum, the American Museum, the Field Museum, the Peabody Museum, where the curators say if they had one intermediate link in their museum they would be happy to show it--but they do not."[6]

I am an advocate of scientific biblical creation and so the evidence I present will be

I Study of preexisting life forms (fossils.)

both scientific and theological. I believe that the God of the Bible created the Earth as described in the Book of Genesis approximately 6,000 years ago. I know a lot of you, including Christians, will consider me daft after that statement, but I think if given the chance I can show that it is not such a stretch to believe that God created as He said.

One possible and plausible scientific explanation for a literal six days of creation is Dr. Russell Humphreys's, *White Hole Cosmology*. The creation could have taken only six days, as recorded in the Book of Genesis, but according to this cosmology, billions of years would have gone by in distant parts of the Universe.[7]

16 "God made two great lights--the greater light to govern the day and the lesser light to govern the night. He also made the stars. 17 God set them in the expanse of the sky to give light on the earth, 18 to govern the day and the night, and to separate light from darkness. And God saw that it was good. 19 And there was evening, and there was morning--the fourth day."

Genesis 1: 16-19

Perhaps now would be a good time to make clear some terms I am using. One definition of intelligent design states, "Intelligent design can be defincd as a theory that holds that 'certain features' of living things were designed by an 'intelligent cause' as opposed to being formed through purely natural means. The intelligent design theory does not name the intelligent cause, and it does not claim that everything is designed, thus allowing for evolution/natural causes to play a role."[8]

Three forms of intelligent design are: biblical creation, creation science and scientific biblical creation. The primary differences between them are:

1.) In biblical creation the Bible is the only evidence required to defend the Bibles' account of creation.

2.) Creation science uses only science to explain the Bibles' account of creation.

3.) Scientific biblical creation uses both the Bible *and* scientific evidence to support the Bibles' account of creation.

As I stated earlier, I am a proponent of number 3, scientific biblical creation. It may surprise you to learn that I do not believe biblical creationism should be taught in our public schools. To force the study of a religious belief is clearly unconstitutional because it would violate the rights of both the students and the teachers.

Frankly, a non-believer is not likely to do a very good job teaching what he does not believe. It is the responsibility of parents to teach their children that the Bible can be trusted in regard to all things including science and history.

Creation science, which is also known as scientific creation, is based on scientific evidence and not the teachings of the Bible. In my opinion, it falls short because it does not show who created or why. Even so, I believe it should be taught in public schools as an alternative point of view to evolution.

Evolutionists argue that intelligent design is religion and not science but in reality the argument is not about science versus theology; it is about one belief versus another belief with science being used to prove one or disprove the other. They have ignored the fact that secular humanism has been declared to be a religion by the United States Supreme Court. The following from EDWARDS v. AGUILLARD, (1987) will show the relevance of that statement. Supreme Court Justice Scalia wrote (from his notes):

"'Senator Keith and his witnesses testified essentially as set forth in the following numbered paragraphs':...

...(5) 'The censorship of creation science has at least two harmful effects. First, it deprives students of knowledge of one of the two scientific explanations for the origin of life, and leads them to believe that evolution is proven fact; thus, their education suffers, and they are wrongly taught that science has proved their religious beliefs false. Second, it violates the Establishment Clause. The United States Supreme Court has held that secular humanism is a religion. Thus, by censoring creation science and instructing students that evolution is fact, public school teachers are now advancing religion in violation of the Establishment Clause.'"[9]

There are thousands of scientists who have rejected Darwinian evolution based on their examination of the scientific evidence.[10] The reason why two scientists who look at the same evidence will draw differing conclusions is quite simple; everyone comes to the table with a different set of beliefs and interprets the evidence based on those beliefs. A perfect example of this is the many millions of fossils that have been discovered around the world. An evolutionary scientist will look and see the remains of animals buried in different layers of earth over a long period of time. A creation scientist will see millions of animals buried quickly by a worldwide flood as described in the Bible's Book of Genesis.

Scientific study normally deals with

processes that can be repeated and observed. Creation cannot be observed, repeated, or tested and as in the example of interpreting the age of fossils, scientific studies relating to creation rely heavily on one's preconceived beliefs.

Evolution has been taught in our schools as fact and is continually being reported as such in the media. Because of this, many Christians have accepted theistic evolution, the belief that God created through evolution, without realizing how harmful this viewpoint can be to their faith.

If science can "prove" that man has evolved and the Earth had its beginning billions of years ago, then the events and times of creation as given in Genesis 1 and 2 cannot be true. And if the Bible cannot be trusted on its first pages then how can any of it be trusted?

Genesis 1 through 11 is the foundation for all the theology and morality that is found in both the Old and the New Testaments. If Genesis can be interpreted to conform with worldly beliefs then why not the entire Bible?

Let me make an appeal to the logic of all Christians who believe that God created through evolution. The Bible tells us that death entered the world because of the sin of the first man, Adam. If he was the product of evolution then there would have been death in the world before Adam. If man has evolved then the Bible is in error and clearly not the Word of God. If a Christian is a follower of Christ, how can they deny what He has said is true?

Jesus referred to Genesis as literal history and not as allegories, fables, or parables when He spoke of the Flood. If He was wrong about the Flood then, "Let us eat and drink, for tomorrow we die," because that's all there is.

Many Christians have compromised their faith by believing in the gap theory which tries to combine evolution and creationism by treating each day of creation as an indefinite period of time. This will not work for various reasons, such as:

1.) Trustworthy Hebrew dictionaries, like the *Brown-Driver-Briggs Hebrew and English Lexicon (BDB)* agree that the Hebrew word for day (yom) as it is used in the first week of creation (in Genesis,) must represent a literal 24 hour day. That is to say, whenever yom is used with a number or with morning or evening it must signify a literal 24 hour day.[11]

2.) God defined and established the "day" as the period of light in the daily cycle of light and dark. Unless you try to read something different into it, the meaning is very clear...

4 "God saw that the light was good, and he

separated the light from the darkness. 5 God called the light 'day,' and the darkness he called 'night.' And there was evening, and there was morning - the first day."

Genesis 1: 4-5

3.) If there had been millions of years of evolution (death and disease) before the sixth day, would God have called that "very good?"

31 "God saw all that he had made, and it was very good. And there was evening, and there was morning - the sixth day."

Genesis 1: 31

4.) God tells us clearly in Genesis that man did not evolve. Adam was the first man and on the sixth day God created him using elements from the earth. And while we're on this subject, Eve did not evolve either; God formed her from one of Adam's ribs. Often people will laugh at this and the Bible because men have the same amount of ribs as women. This is really a silly argument for two reasons. If a man with one leg fathers a child, will that child have only one leg? Also, did you know that human ribs can grow back when they are carefully removed by surgeons and used for reconstructing damaged bones? Ironically, this has been called the "Eve procedure."[12]

7 “the LORD God formed the man from the dust of the ground and breathed into his nostrils the breath of life, and the man became a living being.”

Genesis 2: 7

21 “So the LORD God caused the man to fall into a deep sleep; and while he was sleeping, he took one of the man's ribs and closed up the place with flesh. 22 Then the LORD God made a woman from the rib he had taken out of the man, and he brought her to the man.
23 The man said,

'This is now bone of my bones
and flesh of my flesh;
she shall be called 'woman,'
for she was taken out of man.'"

Genesis 2: 21-23

† † †

You may wonder why many people are so willing to accept evolutionary theory as fact. The answer is simple; if man did not evolve, then he was created. And if he was created, then there is a Creator who man must answer to. There is no other choice.

Some famous men who based their justification of Nazism, Fascism, Communism, and Humanism on evolutionary teachings are, respectively: Adolf Hitler, Benito Mussolini, Karl Marx, and Carl Sagan. The following quotation is by Sir Arthur Keith who was one of the foremost evolutionists and anthropologists[II] of the twentieth century:

"The German Fuhrer, as I have consistently maintained, is an evolutionist; he has consciously sought to make the practice of Germany conform to the theory of evolution."[13]

The following quote by Adolf Hitler is from *Mein Kampf*:

"The pre-requisite for improvement of the species lies not in the union of the superior and the inferior, but in the complete victory of the former. The stronger must dominate and not mix with the weaker, and thereby sacrifice its own greatness. Only the born weakling can feel this to be cruel. He is indeed but a weak and limited creature. If this law did not prevail, any higher evolution of all organic life would be unthinkable."[14]

Evolutionists are quick to deny the connection between Darwin and the atrocities committed by these and other men but consider the following:

II Those who study the history of man.

Karl Marx (the founder of Communism) wrote that *The Origin of the Species* agreed with his communist policies and Stalin believed the atrocities committed under his regime (estimated to be 20 million lives) were justified by Darwin's evolutionary principles.[15]

Mao Tse Tung, a Communist leader who came to power in China in 1949, regarded Charles Darwin and Thomas Huxley[III] as his favorite authors. Leon Trotsky,[IV] who persecuted the Christian church, said, "Darwin stood for me like a mighty doorkeeper at the entrance to the temple of the universe."[16]

Evolution is a lie, the purpose of which is to keep people from accepting the message of the Gospel. Jesus said this Himself in John 5: 46-47. Keep in mind that Genesis was written through Moses:

46 "If you believed Moses, you would believe me, for he wrote about me. 47 But since you do not believe what he wrote, how are you going to

III Thomas Huxley was known as Darwin's Bulldog and authored *Evolution and Ethics* which was published in 1898.

IV Trotsky (1879-1940) was a Russian Marxist and the first leader of the Red Army.

believe what I say?"

John 5: 46-47

If someone doesn't believe the account of history as given in Genesis: the Creation, the Fall, the Flood, then why should they believe in the need for a Savior?

For more than a century we have been told that evolution occurred gradually over millions of years. Now, evolutionists have had to give up this argument because the evidence (transitional fossils) cannot to be found. Some, out of ignorance or deceit, are still claiming the fossil record proves evolution to be fact. The following quote is from a notable evolutionist:

"In a published interview, Dr. Niles Eldredge, an invertebrate paleontologist[V] at the American Museum of Natural History, stated:

'But the smooth transition from one form of life to another which is implied in the theory is... not borne out by the facts. The search for 'missing links' between various living creatures, like humans and apes, is probably fruitless... because they probably never existed as distinct transitional types... But no one has yet found any evidence of such transitional creatures. This oddity has been attributed to gaps in the fossil

V Someone who studies fossils of creatures without a backbone.

record which gradualists expected to fill when rock strata of the proper age had been found. In the last decade, however, geologists have found rock layers of all divisions of the last 500 million years and no transitional forms were contained in them. If it is not the fossil record which is incomplete then it must be the theory.'"[17]

Because there is no scientific evidence for gradual evolution, new and even more fanciful theories have been developed to overcome this problem. A theory that you may be familiar with is punctuated equilibrium. A dinosaur laid an egg, and a dinosaur crawled out. A dinosaur laid an egg and... a bird flew out! This would certainly explain the absence of any transitional fossils.

Another popular theory (especially in Hollywood) is directed panspermia in which supposed aliens "planted" their seed on Earth. To that we should ask, "How did this alien race begin?" and "Could that alien race have evolved somewhere else in the Universe?"

Actually, a more relevant question is, could life have evolved here or anywhere else in the Universe as evolutionists claim by spontaneous generation? (the theory that living organisms can develop in non-living matter independently of other living matter.) The answer is an emphatic no.

Spontaneous generation was disproved by Redi (1688,) Spallanzanu (1780,) Virchow (1858,) and Pasteur (1860.) The works of these men and others have established the Law of Biogenesis; life comes only from preexisting life and will only perpetuate its own kind.

24 "And God said, 'Let the land produce living
creatures according to their kinds: the
livestock, the creatures that move along the
ground, and the wild animals, each according to
its kind."
And it was so.'"

Genesis 1: 24

Francis Crick was an atheist and one of the world's most prestigious biologists and codiscoverer of DNA, the most complex molecule known to man (1953.) He examined the possibility of life coming into existence by chance on Earth and stated that it appears to be "almost a miracle."[18]

Another atheist, Sir Fred Hoyle of Cambridge University, was one of the most noted astronomers and mathematicians of the twentieth century. He examined the probabilities of life forming on this planet; the following quotation is attributed to him:

"The notion that not only the biopolymer

but the operating program of a living cell could be arrived at by chance in a primordial organic soup here on the Earth is evidently nonsense of a high order."[19]

I have the book, *Genesis On Planet Earth*, by William Day. The only biographical information I found in it states that Mr. Day was a research associate in the department of chemistry at the University of Iowa. From what I have read I have to believe it was written from an evolutionary point of view. The author wrote the following:

"A curious flaw of human nature is to permit the imagery of a catchy phrase to shape one's reasoning. Haldane's hot dilute soup became 'primordial soup,' a feature that has been popularized for nearly 50 years without geologic evidence that it ever existed."[20]

Hoyle also examined the probability of life arising on some other planet in the entire assumed age of the Universe (which is currently believed to be approximately 13.77 billion years)[21] and concluded that it could not have happened:

"The probability of an evolutionary origin of life was essentially zero—no greater than the probability that a tornado sweeping through a junkyard would assemble a Boeing 747."[22]

I hope you will bear with me while I try to show evidence for a young Earth. I do not know the age of the Earth, but it is certainly closer to 6 thousand years than 4.5 billion years (the estimated age of our solar system.)[23] The evidence to support this claim is very powerful to those who are willing to open their mind to what God was saying to evolutionists on May 18, 1980.

The different layers of the Earth's surface are known as strata and evolutionists use this as a way of dating living things by where their fossils are found. It is believed that the lower the level of strata that a fossil is found in, the further back in time that it lived.

A good example of this is the Grand Canyon. The many layers of strata that are visible to study are explained by evolutionary geologists as having been caused by erosion over millions of years by the Colorado River. Stay with me here; the point I am trying to make is not that the different layers of strata are easily seen. The point is that those different layers are only supposed to develop over millions of years.

On May 18, 1980, Mt. St. Helens in Washington erupted and the events that followed changed the thinking of many geologists. The volcanic eruption produced many layers of strata to a depth of 600 feet. The rapid formation caused a 140 foot deep canyon through the strata and created a 1/40th scale model of the Grand Canyon. What was supposed to take millions of years to occur took only days.

Evolutionists claim that the age of the strata can be determined by the fossils that are found in it, and the age of the fossil can be determined by the layer of strata that it is found in. Did you catch that?

Dating methods used by evolutionists are unreliable because they are based on the assumption that the Earth is billions of years old.[VI] Lava from Mt. St. Helens was tested by the potassium-argon method fourteen years after the eruption and dated at 350,000 years.[24] Likewise, 50 year old lava from Mt. Ngauruhoe[VII] was dated at 3.5 million years.[25]

VI For a detailed explanation of this see: <https://answersingenesis.org/geology/radiometric-dating/radiometric-dating-problems-with-the-assumptions/#fn_1>.

VII Mount Ngauruhoe is an active volcano in New Zealand. As a side note, it was used as a stand-in for the fictional Mount Doom in the *Lord of the Rings* trilogy.

4 “Where were you when I laid the earth's
foundation?
Tell me, if you understand.”
Job 38: 4

Through the eruption of Mt. St. Helens, God has shown us how strata and fossils were formed, by what scientists refer to as a global hydraulic cataclysm. This event is described in the Bible as the Genesis Flood. Dead animals either decay or are devoured by insects or other animals. The billions of fossils perfectly preserved in the Earth's strata are a record of sudden mass death and destruction caused by Noah's Flood. The global flood carried the sand, silt, clay, plants and animals that produced the layers of rock which contain these fossils. Keep in mind that most fossils are found in sedimentary rock which is formed by... flowing water.

Ask an evolutionist for a reasonable explanation of how fossils are found extending through one layer of strata to another if the layers took thousands of years to form. It can't happen. How could the lower part of the fossil have been preserved as the strata formed around it over thousands of years? How could the part that stuck out of the first layer not have decayed, been eaten, or broken off while the next layer formed around it over thousands of years? As I said, it can't happen and it isn't happening today, which is a real problem for evolutionists.

† † †

It is bad science to ignore the clear meaning of facts because they do not conform to the theory of evolution or the various Big Bang theories which assume that the Universe created itself. Many facts are ignored for this reason. Let me list just a few examples that illustrate how scientific evidence which contradicts an old Earth will not be accepted by evolutionists:

The Earth's magnetic field has been accurately measured since 1835 and has been shown to be decaying at a half life of 1,400 years, meaning that 1,400 years ago it was twice as strong as it is now. At that rate just 10,000 years ago the magnetic field of the Earth would have been strong enough to melt the planet.[VIII]

Each time comets orbit the Sun they lose some of their mass and it has been determined that they would totally dissipate in about 10,000 years. Comets are believed to be the same age as the Solar System and no practical source has

---

VIII For technical details see... <https://cdn-assets.answersingenesis.org/doc/articles/am/v7/n4/ECP-p873-877-Earths-Magnetic-Field.pdf>.

been found for the creation of new comets. Notice that I said, "no practical source," because there are theories such as the Oort Cloud which is, "A theoretical group of planetary leftovers that supposedly orbits the solar system... But, no confirmed direct observations of the Oort Cloud have been made."[26] That sounds like another attempt by evolutionists to explain what can't be explained if the Earth is millions of years old.

Fossils of salamanders have been found in layers of strata which are alleged to be 150 million years old.[27] The problem for evolutionists is that the fossils and the salamanders walking around today are identical. There has been no change. And salamanders are not unique in this sense because there are many other examples of "living fossils" today.

Evolutionary scientists believe that all dinosaurs became extinct at least 65 million years ago, but recently Tyrannosaurus rex bones were discovered that contained soft tissue including blood vessels, cells, and connective tissue which could not have remained intact for millions of years.[28]

The following are some great resources for anyone seeking information about the creation vs. evolution debate:

Institute for Creation Research at:
icr.org,
Creation Ministries International at:
creation.com, and
Answers in Genesis at:
answersingenesis.org.

† † †

The following quote regarding the First and Second Laws of Thermodynamics is by Dr. Henry M. Morris and Dr. John D. Morris. Keep in mind that by "First Cause" they undoubtedly are referring to God:

"Thus, the Second Law proves, *as certainly as science can prove anything whatever*, that the universe had a beginning. Similarly, the First Law shows that the universe could not have begun itself. The total quantity of energy in the universe is a constant, but the quantity of *available* energy is decreasing. Therefore, as we go *backward* in time, the available energy would have been progressively greater until, finally, we would reach the beginning point, where available energy equaled total energy. Time could go back no further than this. At this point both energy and time must have come into existence. Since energy could not create itself, the most scientific and logical conclusion to which we could possibly come is that the First Cause created energy (including matter), as well as time.'"[29]

Another way of stating the above would be, "In the beginning, God created the Heavens and the Earth." The following quotations on the subject of information are from *In the Beginning was Information* by Dr. Werner Gitt:

"Because information is required of all life processes, it can be stated unequivocally that information is an essential characteristic of all life. All efforts to explain life processes in terms of physics and chemistry only, will always be unsuccessful. This is the fundamental problem confronting present-day biology, which is based on evolution."[30]

"Every piece of creative information represents some mental effort and can be traced to a personal idea-giver who exercised his own free will, and who is endowed with an intelligent mind."[31]

"There is no known law of nature through which matter can give rise to information, neither is any physical process or material phenomenon known that can do this."[32]

In 1802, theologian William Paley wrote that if you find a pocket watch in a field, the most reasonable conclusion is that someone dropped it there and not that it was created there by any natural forces. He then argued that other even more complex living things could not have come about by natural forces and therefore must have been created.

A watch requires a watchmaker and so intelligent design requires a designer. The Universe and all it contains are the work of an intelligent designer. Man could not have evolved and is therefore the product of intelligence. God is the Watchmaker. God is the Intelligent Designer. And He has revealed himself to us by stepping into His creation in the form of a man, Jesus Christ.

1 "In the beginning God created the heavens
and the earth. 2 Now the earth was
formless and empty, darkness was over
the surface of the deep, and the Spirit of
God was hovering over the waters.

3 And God said, 'Let there be light,' and there
was light. 4 God saw that the light was
good, and he separated the light from the
darkness. 5 God called the light 'day,' the
and the darkness he called 'night.'
And there was evening, and there was
morning--the first day."

Genesis 1: 1-5

1 “In the beginning was the Word, and the Word
was with God, and the Word was God. 2 He was
with God in the beginning.
3 Through him all things were made; without
him nothing was made that has been made. 4 In
him was life, and that life was the light of men.”

John 1: 1-4

Just as the Bible can be trusted regarding creation, it can also be relied on concerning salvation:

11 “And this is the testimony: God has given us
eternal life, and this life is in his Son. 12 He who
has the Son has life; he who does not have the
Son of God does not have life.

13 I write these things to you who believe in the
name of the Son of God so that you may know
that you have eternal life.”

1 John 5: 11-13

† † †

Let's take another look at what Darwin had to say about his theory:

"Why then is not every geological formation and every stratum full of such intermediate links? Geology assuredly does not reveal any such finely graduated organic chain; and this, perhaps, is the most obvious and gravest objection which can be urged against my theory."

Well Charles, your hopes have been dashed and the foundation of your theory has crumbled; there are no transitional fossils. Had you been proven right, millions of the fossils that have been uncovered would have characteristics of fossils that came before them and those that appeared after them. There are no such fossils.

If Charles Darwin was alive today and able to see the multitude of evidence that has accumulated against his theory, would he be willing to face the alternative, that there *must* be a creator God? Would he admit that the biblical account of creation is indeed trustworthy? I doubt it.

For Charles Darwin, and perhaps millions of others with the same mind set, that would be a pill too hard to swallow. It is the theory of evolution that frees them from the authority of God's Word.

THE GOSPEL OF JESUS CHRIST

6 "I am astonished that you are so quickly deserting him who called you by the grace of Christ and turning to a different gospel—7 which really is no gospel at all. Evidently some people are throwing you into confusion and are trying to pervert the gospel of Christ."

Galatians 1: 6-7

THE GOSPEL OF JESUS CHRIST

It would be easy to make this short and ask you to choose, as we all must do, to either accept the Gospel of Jesus Christ or reject it. However, I believe the desire to make all things quick and effortless is one of the problems with the church today. Since it is how a person responds to the Gospel that will determine their eternal destiny, let's not rush through this.

Many people are not being presented with the whole Gospel, and as the Apostle Paul relates in the Book of Galatians, anything but the actual Gospel is no Gospel at all.

6 "I am astonished that you are so quickly deserting him who called you by the grace of Christ and turning to a different gospel—7 which really is no gospel at all. Evidently some people are throwing you into confusion and are trying to pervert the gospel of Christ."

Galatians 1: 6-7

Before I can even begin to present you with the Gospel, there are some fences that may need to be torn down. Satan is known by many names, a few of those are: Beelzebub, Deceiver, the Devil, and the Accuser. Although some people don't believe that he is actually a living being, the Bible tells us that he is. Jesus gave us a pretty good description of him, as recorded in the New Testament Book of John.

44 "... He was a murderer from the beginning, not holding to the truth, for there is no truth in him. When he lies, he speaks his native language, for he is a liar and the father of lies. "

John 8: 44

Perhaps the biggest lie that Satan ever told was when he convinced Adam and Eve that they could be like God. And, the "god of this age" doesn't seem to have mellowed over time, because surely he must be the one behind the current deceptions being taught in many of our public schools. I am referring to the false teachings regarding tolerance and truth. I think I can address these topics together because, in some ways, they relate to each other.

One definition of tolerate is, "To allow the existence, presence, practice, or act without prohibition or hindrance."[33] This would certainly be in agreement with the First Amendment to the

Constitution as it relates to our God given right to free speech.

It is the politically correct definition of tolerance, that all values, beliefs, and truths are equal, that is being used to suppress biblical positions on morality and law. Christians who speak out against issues such as gay marriage are accused of being intolerant. But if all beliefs are equal, then aren't the politically correct showing themselves to be intolerant of Christian beliefs?

This same type of hypocrisy has entered the church and may be the major cause of the watered down, non-offensive Gospel we see in many Christian churches today. The Gospel is viewed as intolerant because Jesus said that *He* was the gate, and *He* was the door. But if He is who He said He is then He certainly has the right to say:

6 "Jesus answered, 'I am the way and the truth and the life. No one comes to the Father except through me.'"

John 14: 6

Is truth absolute or relative (objective or subjective?) Can all truths be equal? Have you ever said, "That may be your truth, but it is not my truth?" I hope not because truth can be either absolute or relative. Without a doubt there are absolute truths: 2+2 always equals 4, there are no round squares or square circles, 10 is always more than 9. All "truths" are not equal.

I need to elaborate on that a little. Absolute truth is something that is unalterable, inflexible and always true. A relative truth is something that is changeable and not always true, but a person's perception of something does not make it true. Let me back up. The problem is not that there are *either* absolute truths *or* relative truths; the problem arises in differentiating between the two.

Relative truths are, more appropriately, personal opinions or beliefs, such as, "blue is the best color," and therefore can be "my truth but not your truth." It is what someone believes to be true at any given time. But, as in the example above, this truth could change tomorrow if that person should decide in favor of another color.

The trouble is that postmodern relativism has attempted to erase the line between relative and absolute truth and declare that there are no moral absolutes. Few rational people will argue that there is no such thing as absolute truth, but political correctness dictates that there are no absolute truths concerning morality, God and religion.

Jesus said there is only one God and only one way to avoid eternal damnation (His truth.)

Others tell us that there is more than one way (their truth.) Therefore, either He is telling the truth or they are. It *cannot* logically be both. There cannot be only one way *and* many ways. So, if someone professes to be a Christian and yet believes, as more than six in ten "Christians" do, that Jesus is not the only way,[34] are they, in all honesty, Christian? If anyone has truly put their faith in Him for salvation, how can they deny what He has said to be true? If He was wrong about *anything*, He is not who He said He is and will not be able to save them.

Often people reject the teachings of the Bible to decide for themselves what is moral and necessary to be saved. Have you noticed that when they do, they always seem to use themselves as the standard? But God doesn't "grade on a curve."

23 "for all have sinned and fall short of the glory of God, "

Romans 3: 23

Tolerance and relative truth are interfering with people accepting the Gospel of Jesus Christ because they can lead them to believe that God is whomever they want Him to be.

16 "You turn things upside down, as if the potter were thought to be like the clay! Shall what is

formed say to him who formed it, "He did not make me"? Can the pot say of the potter, "He knows nothing"? "

Isaiah 29: 16

Like all lies, they don't stand up to close examination. Even the logic is faulty. When someone claims a Christian is being intolerant for expressing biblical values, they are showing *themselves* to be intolerant and contradicting *their own belief* that all beliefs are equal. If someone states that there is no such thing as absolute truth one need only reply, "Are you *absolutely* sure?" to show how absurd and irrational their argument is. By stating there is no such thing as absolute truth they are making an absolute statement that there are no absolute truths, which you can see (if you have followed this!) is a self-defeating and illogical argument.

† † †

There are other things that can get in the way of someone believing in the Gospel. The theory of evolution and the various "Big Bang" theories serve as an attempt to prove there is no need to believe in a creator and thus show that the Bible cannot be trusted.

The reason these fences need to be torn down is, if you don't believe that God created the

Universe as He said in the Bible, and you are hampered by false teachings such as evolution and relativism, it can interfere with your belief in the need for a savior.

All the theology that is found in the New Testament has its foundation in the first eleven chapters of Genesis. Until the advent of Darwin's theory of evolution, the Book of Genesis was almost unanimously believed to be historical narrative. But now, to conform with evolutionary teachings, Genesis has been reduced by many to a collection of allegorical tales.

† † †

I will now do my best to present the Gospel of Jesus Christ and we will start, not in the New Testament, but where it begins, in the Bible's first book, Genesis.

God told Adam and Eve that if they ate the fruit from the tree of the knowledge of good and evil, they would die. After they were deceived by the serpent, who was under Satan's influence, God brought judgment on Satan, Adam, and Eve. He cursed Satan and the ground that man worked. It was the sin (disobeying God by eating from the tree of the knowledge of good and evil) of the first man Adam that brought both physical and spiritual death into the world, as God had promised it would. All humans are descendants of Adam and Eve and have inherited their sinful nature. But, if you bear with me, I will show that God did not abandon His creation.

The following scripture is known as the Protoevangelium, the first announcement of the Gospel. It tells us that although Satan would harm Jesus (*bruise his heel,)* Jesus, the "seed" of the woman, would defeat Satan (*bruise his head.)*

14 “So the Lord God said to the serpent, "Because you have done this,

"Cursed are you above all the livestock
and all the wild animals!
You will crawl on your belly and you will eat
dust all the days of your life.

15 And I will put enmity between
you and the woman,
and between your offspring and hers;
he will crush your head,
and you will strike his heel.'"

Genesis 3: 14-15

The New Testament Book of Luke gives a detailed genealogy from Adam to Jesus to show that He is the "seed" of the woman. The Old

Testament was written in Hebrew and the Hebrew word for Adam is the same as the word for man. Adam, the first man, brought death into the world. Jesus, who is both man and God, was a physical descendant of Adam. If these were not actual events there would be no reason for the Bible to give us a genealogy.

The Bible tells us that Jesus, who referred to himself as the "Son of Man," is the "last" Adam (1 Corinthians 15: 45.) Adam (man) brought sin into the world, and Jesus, the descendant of Adam (Son of Man) came to redeem the world.

4 "But when the time had fully come, God sent his Son, born of a woman, born under law, 5 to redeem those under law, that we might receive the full rights of sons."

Galatians 4: 4-5

† † †

Now, let's take a quick look at the ancient Hebrew Tabernacle and the Temples which are historically verifiable. The Tabernacle was portable, contained in a tent, and was constructed in the time of Moses (app. 1450 B.C.) King Solomon's Temple was completed in 960 B.C. and was destroyed in 586 B.C. by Nebuchadnezzar, King of Babylon. The second temple was started in 516 B.C. King Herod resumed work on the temple in 20-19 B.C. and work was not completed until 62-64 A.D. Jesus predicted it's destruction (Mark 13: 1-3) and that occurred in the year 70 A.D. by Roman troops under the command of general Titus who later became Caesar.

The importance of the Old Testament Tabernacle and the Temples is that they were the center of the sacrificial system that God had established for the atonement of sin. The Bible tells us that there can be no forgiveness of sin without the shedding of blood.

22 "In fact, the law requires that nearly everything be cleansed with blood, and without the shedding of blood there is no forgiveness."

Hebrews 9: 22

And we learn in the Book of Numbers that the sacrificial lambs had to be perfect.

19 "Present to the Lord an offering made by fire, a burnt offering of two young bulls, one ram and seven male lambs a year old, all without defect."

Numbers 28: 19

This requirement is important, which you will see as we move on to the time of the New Testament.

29 "The next day John saw Jesus coming toward

him and said, 'Look, the Lamb of God, who takes away the sin of the world!'"

John 1: 29

John the Baptist announced that Jesus was the Messiah, which is Hebrew for "Anointed" or "Savior." It is translated as Christos in Greek and Christ in English. How can we know that Jesus, the "Lamb of God," is the Messiah? One way is the prophecy given in Isaiah 53 that describes Him. I hope you will read the entire chapter because it is an incredible prophesy that describes Jesus and events that occurred in His life. Here it is in part:

7 "He was oppressed and afflicted,
yet he did not open his mouth;
he was led like a lamb to the slaughter,
and as a sheep before her shearers is
silent,
so he did not open his mouth."

Isaiah 53: 7

Jesus is the unblemished lamb. He is the only one who has ever lived a perfect, sinless life and is therefore the only one whose sacrifice could pay the sin debt of the world.

10 "And by that will, we have been made holy through the sacrifice of the body of Jesus Christ

once for all."

Hebrews 10: 10

In the Book of John we see that Jesus even asked His enemies, the Pharisees, if they could convict Him of any sin, and they could not.

45 "Yet because I tell the truth, you do not believe me! 46 Can any of you prove me guilty of sin? If I am telling the truth, why don't you believe me?"

John 8: 45-46

† † †

Every religion that I am aware of, except Biblical Christianity, teaches some system of works. Roman Catholicism denies this. They claim that God provides the grace to perform the works they believe are necessary for salvation. However, their arguments cannot stand up to scripture. Do you believe that your own good life, your good works, will earn you eternal life? If that is your "truth," does it agree with God's Truth?

4 "You who are trying to be justified by law have been alienated from Christ; you have fallen away from grace."

Galatians 5: 4

28 “Then they asked him, 'What must we do to do the works God requires?'

29 Jesus answered, 'The work of God is this: to believe in the one he has sent.'"

John 6: 28-29

The reply Jesus gave is very significant because He changed the plural "works" from the question to the singular "work" in His answer. The only acceptable work is to believe in Jesus, and God does the work for us, "This is the work of God."

Before I attempt to (finally?) spell out what the Gospel is, let me also say that it is not what I referred to earlier as a, "watered down gospel." The Gospel is more than John 3: 16.

16 "For God so loved the world that he gave his one and only Son, that whoever believes in him shall not perish but have eternal life.”

John 3: 16

The Book of James tells us that knowing the truth about God has no redeeming quality. An intellectual knowledge of the Gospel is not the same as the personal relationship that comes with its acceptance.

19 “Show me your faith without deeds, and I will show you my faith by what I do. You believe that

there is one God. Good! Even the demons believe that--and shudder."

James 2: 19

Theologian Walter Scott (1796-1861) put it very well when he said that the Gospel of Jesus Christ, that was taught by the apostles and can be found in the Bible, consists of "facts to be believed, commands to be obeyed, and promises to be anticipated." We need to believe that Jesus is the Son of God who became man to pay the debt for our sins. That he was crucified, died, was buried, and rose again on the third day. We must admit that we truly are sinners and that we can do nothing to restore our broken relationship with the Lord other than accept His free gift of salvation that He paid the price for.

Here is the part that is too hard for many people. We must ask Jesus into our lives and declare Him Lord. What do we receive with this? God promises to send His Spirit to indwell our hearts and to guide us. He promises that all our sins are forgiven: past, present and future. We also have the assurance that at the Resurrection of the dead we will be given a new spiritual body and that we will never die. That is the Gospel, the "Good News" of Jesus Christ.

† † †

What I have written is not about me or for me, it is about Jesus and it is for you. But perhaps it will be helpful if I reveal some things about my life that may relate to your own.

It was not that long ago, but it seems like forever to think back on the time before I began my walk with the Lord. Twelve years of Catholic school had not really answered the questions we all ask, at least not on a personal level. What is God really like, and what is His purpose for me? What will become of me when I die? If there is a God, and someday I stand in His court, will my opinion of my life have any bearing on His decision? Will I be "good" enough to be saved?

3 "All have turned aside, they have together become corrupt; there is no one who does good, not even one."

Psalm 14: 3

I determined to find the answers. Like many Catholics, I was not very knowledgeable about God's Word. I had the old *Douay-Rheims Version* of the Bible, and like the *King James Version*, I found it difficult to read. I bought a *New International Version* (NIV) 1984 edition study Bible and it was like a breath of fresh air to be able to read in modern language the Word of God as He spoke to my heart.

One night as I was listening to Dr. D. James Kennedy from Coral Ridge Ministries on television, he gave an invitation to accept the Gospel. I asked Jesus into my life. It was like a weight had been lifted from my shoulders and from that point on, the Lord has been guiding me and opening up His Truth through the pages of the Bible, as He has for so many before me.

Hard as I tried to remain a Roman Catholic, the more I learned, the more I realized I had to choose between being a Catholic or a Christian. The two beliefs are mutually exclusive; you cannot be saved by both faith *and* works. The only acceptable work was done for us on the cross.

6 "And if by grace, then it is no longer by works; if it were, grace would no longer be grace."
Romans 11: 6

Allow me to paraphrase Elijah's words from 1 Kings 18: 21...

How long will you straddle the issue? If the Bible is the infallible Word of God, then trust it; if the teachings of the Roman Catholic Church are infallible, then trust them.

When I was in Catholic school we were taught that the Protestants from the sixteenth century were heretics who attacked the church. In reality, the Reformers were Catholics who tried to bring the church back to the Gospel of salvation by faith in Jesus Christ, and to re-establish the authority of God's Word. Many of them paid with their lives for printing and distributing Bibles, so we might have the Word of God that had been kept hidden for centuries.

I knew I had to reach others who also needed to be freed from the bondage of salvation through works. To be, as the Apostle Paul states, in Christ, is to be free from a religion that always keeps salvation just out of reach... one more novena, one more rosary, *how much is necessary to pay your sin debt*? The Bible tells us that if we are born again, by accepting the Gospel of Jesus Christ, our salvation is assured. Could anything show this more clearly than the following scriptures?

10 "Anyone who believes in the Son of God has
this testimony in his heart. Anyone who does not
believe God has made him out to be a liar,
because he has not believed the testimony God
has given about his Son. 11 And this is the testi-
mony: God has given us eternal life, and this life
is in his Son. 12 He who has the Son has life; he
who does not have the Son of God does not have
life.
13 I write these things to you who believe in the

name of the Son of God so that you may know that you have eternal life."

1 John 5: 10-13

24 "I tell you the truth, whoever hears my word and believes him who sent me has eternal life and will not be condemned; he has crossed over from death to life."

John 5: 24

If you have not accepted Jesus as your Lord and Savior, then what better time than now?

2 "For he says, 'In the time of my favor I heard you, and in the day of salvation I helped you.' I tell you, now is the time of God's favor, now is the day of salvation."

2 Corinthians 6: 2

All it takes is a simple prayer from the heart to enter into a personal relationship with the One who knew you before time began. Something like this...

Lord, I confess that I am a sinner and that I cannot earn my salvation. I believe that Jesus died on the cross for my sins and that on the third day He rose from the dead. I accept Jesus as my Lord and Savior and I ask you to bring me the peace and assurance that comes from knowing that I have been born again. I ask this in Jesus's name. Amen.

29 “Take my yoke upon you and learn from me, for I am gentle and humble in heart, and you will find rest for your souls. 30 For my yoke is easy and my burden is light.”

Matthew 11: 29-30

7 "Ask and it will be given to you; seek and you will find; knock and the door will be opened to you. 8 For everyone who asks receives; he who seeks finds; and to him who knocks, the door will be opened.”

Matthew 7: 7-8

About the Author

This is the author's second book that relates to the theory of evolution and the reliability of the Bible, but first came years of research and writing including a website that has had almost 400,000 hits.

Many of those were from China where people are hungry for information that has been kept from them. Unfortunately all that ended when China shut the door by blocking the website.

He has written this new book and built a new website, kingandlord.net, which will be used to reach a world that badly needs to know the truth; are we here by chance or is there a Creator?

His latest book, *The Bible Or The Roman Catholic Church* can be found at Amazon.com and other fine sources.

A new book, *Answers in Scripture*, is coming soon!

Robert Rani is a Christian, a father and grandfather, and makes his living by building custom cabinets, moldings, and furniture.

We welcome your feedback at...
robert.rani@att.net

1 Charles Darwin, *On The Origin of Species*, 1859, Chapter 9: On the Imperfection of the Geological Record.

2 Dr. John F. Ashton, *Evolution Impossible: 12 Reasons Why Evolution Cannot Explain the Origin of Life on Earth*, First Edition, 2012, Master Books, A Division of the New Leaf Publishing Group, Inc., Endnote 30: Carl Warner, *The Quest for an Answer, Evolution: The Grand Experiment*, Volume 1, (Green Forest, AR: New Leaf Press, 2009,) p. 77.

3 Chuck Colson, *The Archaeoraptor Fraud*, Breaking Point, accessed November 22, 2015, <https://www.breakpoint.org/bpcommentaries/breakpoint-commentaries-search/entry/13/12450>.

4 Sarah Pruitt, *Piltdown Man Hoax, 100 Years Ago*, History in the Headlines, accessed November 22, 2015, <http://www.history.com/news/piltdown-man-hoax-100-years-ago>.

5 Kenneth Ham, *Creation Ministry in the United Kingdom*, Institute for Creation Research, accessed November 22, 2015, <http://www.icr.org/article/creation-ministry-united-kingdom/>.

6 D. James Kennedy, A.B., M. Div., M. Th., D.D., D. Sac. Lit., Ph. D., Litt. D., D. Sac. Theol., D. Humane Let., From printed version of the sermon, *Creationism: Science or Religion?*, p. 10, Coral Ridge Ministries, P.O. Box 40, Fort Lauderdale, Florida 33302, Used by permission.

7 Creation Wiki, *Encyclopedia of Creation Science*, accessed November 11, 2015, <http://creationwiki.org/White_hole_cosmology>.

8 Dr. Georgia Purdum, *The Intelligent Design Movement, Does the identity of the Creator really matter?*, Answers in Genesis, accessed November

22, 2015,
<https://answersingenesis.org/intelligent-design/is-the-intelligent-design movement-christian/>.

9 Cornell University Law School, Legal Information Institute, 482 U.S. 578, *Edwards v. Aguillard* (No. 85-1513), Argued: December 10, 1986, Decided: June 19, 1987, accessed January 3, 2015,
<http://www.law.cornell.edu/supremecourt/text/482/578#writing-USSC_CR_0482_0578_ZD>.

10 David Cloud, *How Many Scientists Reject Darwinian Evolution?*, March 25, 2015, Fundamental Baptist Information Service, P.O. Box 610368, Port Huron, MI 48061, 866-295-4143, fbns@wayoflife.org, Way of Life Literature, Inc., accessed November 22, 2015,
<http://www.wayoflife.org/index_files/how_many_scientists_reject_darwinism.html>.

11 Ken Ham, *The Necessity for Believing in Six Literal Days,* December 1, 1995, Answers in Genesis, accessed November 22, 2015,
<https://answersingenesis.org/why-does-creation-matter/the-necessity-for-believing-in-six-literal-days/>.

12 *Rib Thoracoplasty,* National Scoliosis Foundation, accessed November 22, 2015,
<http://www.scoliosis.org/resources/medicalupdates/ribthoracoplasty.php>.

13 Sir Arthur Keith, *EVOLUTION AND ETHICS*, G. P. Putnam's Sons, New York, VAN REES PRESS NEW YORK, Chapter 40, An Evolutionary Interpretation of the Second World War, p. 230, [No copyright renewal found. Text is now in the public domain under U.S copyright laws and may be distributed freely.,] accessed October 12, 2013,

<http://www.bearfabrique.org/Evolution08/evolution-and-ethics.html>.
14 Adolf Hitler, *Mein Kampf,* Volume 1: A Retrospect, Chapter XI, Race and People, p. 174, HURST AND BLACKETT LTD., Publishers since 1812, LONDON "NEW YORK" MELBOURNE, This translation of the unexpurgated edition of "Mein Kampf" which was first published on March 21st, 1939.
15 *MARXISM AND NATURAL SELECTION*, Natural Selection and Darwinism, (Based on the works of Harun Yaha,) accessed February 21, 2016, <http://www.naturalselectionanddarwinism.com/marxism.html>.
16 Barry Woolley, *The Darwin/Trotsky Connection*, Creation Ministries International, Creation.com, Originally published in Creation 23,(2,) (March, 2001): 54–55, accessed October 11, 2013, <http://creation.com/darwin-trotsky-connection>.
17 Walt Brown, Ph.D., *In the Beginning: Compelling Evidence for the Creation and the Flood*, Seventh edition, 2001 by Walt Brown, All rights reserved, ISBN 1-878026-08-9, printed in Hong Kong, page 52, quoting *Missing, Believed Nonexistent*, Manchester Guardian (The Washington Post Weekly,) Vol. 119, No. 22, 26 November 1978, p. 1.
18 Crick, F., *Life Itself,* New York: Simon and Schuster, 1981, p. 88.
19 Greg Demme and Jonathan Sarfati, *Big bang critic dies (Fred Hoyle,)* Creation Ministries International, accessed November 22, 2015, <http://creation.com/big-bang-critic-dies-fred-hoyle>.
20 William Day, *Genesis On Planet Earth, The*

Search for Life's Beginning, Second Edition, 1984, Yale University Press, New Haven and London, p. 151.

21 Don Batten, *Age of the earth 101 evidences for a young age of the earth and the universe*, Published: 4 June 2009, Creation Ministries International, accessed November 23, 2015, <http://creation.com/age-of-the-earth>.

22 Duane Gish, Ph.D., *The Arkansas Decision on Creation-Science*, Institute for Creation Research, accessed November 23, 2015, <http://www.icr.org/article/arkansas-decision-creation-science/>.

23 Don Batten, *Age of the earth 101 evidences for a young age of the earth and the universe*, Published: 4 June 2009, Creation Ministries International, accessed November 23, 2015, <http://creation.com/age-of-the-earth>.

24 S. A. Austin, "*Excess Argon within Mineral Concentrates from the New Dacite Lava Dome at Mount St. Helens Volcano*," Creation Ex Nihilo Technical Journal 10.3 (1996): 335–343, this information was retrieved from answersingenesis.org, October 4, 2015, <https://answersingenesis.org/geology/mount-st-helens/argon-in-mineral-concentrates-from-mount-st-helens-volcano/>.

25 Dr. Andrew A. Snelling, "*Radioactive 'Dating' Failure Recent New Zealand Lava Flows Yield 'Ages' of Millions of Years*," Originally published in *Creation* 22, no 1 (December 1999): 18-21, accessed from answersingenesis.org, December 27, 2015, <https://answersingenesis.org/geology/carbon-14/radioactive-dating-failure/>.

26 Brian Thomas, M.S., *New Comet Origins Idea*

Adds New Problems, Institute for Creation Research, accessed December 5, 2015, <http://www.icr.org/article/new-comet-origins/>.

27 John Roach for National Geographic News, *Salamander Origins Pegged To Asia*, March 28, 2001, accessed December 28, 2015, <http://news.nationalgeographic.com/news/2001/03/0328_salamander.html>.

28 Dr. David Menton, *#3 Soft Tissue in Fossils 10 Best Evidences From Science That Confirm a Young Earth*, October 1, 2012, accessed from answersingenesis.org December 28, 2015, <https://answersingenesis.org/fossils/3-soft-tissue-in-fossils/>.

29 Dr. Henry M. Morris, Dr. John D. Morris, *The Modern Creation Trilogy, Science & Creation, Volume II*, p. 132, First printing: November 1996, Copyright 1996 by Master Books, All rights reserved, Printed in the United States of America, No part of this book may be used or reproduced in any manner whatsoever without written permission of the publisher, except in the case of brief quotations in articles and reviews, For information write: Master Books, Inc., P.O. Box 727, Green Forest, AR 72638, ISBN: 0-89051-221-3, Used by permission.

30 Dr. Werner Gitt, *In the Beginning was Information*, 2nd English Translation, 2000, CLV Christliche Literatur-Verbreitung e. V., Translation to English: Prof. Dr. Jaap Kies, South Africa, p. 9, accessed November 29, 2013, <http://www.clv-server.de/pdf/255255.pdf>, Used by permission.

31 Ibid, p. 113.

32 Ibid, p. 79.

33 "tolerate." Dictionary.com Unabridged, Random

House, Inc. 07 Feb. 2015, accessed April 5, 2014, <http://dictionary.reference.com/browse/tolerate>.

34 Audrey Barrick, Christian Post Reporter, Christian Post, CP Politics, *Many Americans Say Christianity Not the Only Way to Eternal Life*, December 19, 2008, accessed 1/6/2015, <http://www.christianpost.com/news/many-americans-say-christianity-not-the-only-way-to-eternal-life-36023/>.

Made in United States
Cleveland, OH
05 November 2026

47666205R00039